PERENNIAL MATH

Tournament Tests

Grades 5-6

PERENNIAL MATH TOURNAMENT TESTS

These are a collection of tests from past Perennial Math Tournaments. Each tournament has two rounds of testing. Each student takes an Individual Test during the first round and then students combine with their teammates to take a Team Test during the second round. The solutions included in this book are only a short sample of the many possible solutions that could be used to solve each problem.

INDIVIDUAL TESTS GRADES 5-6

30 minute practice tests for individual students

Grade 5-6 Individual Event (30 Minutes)

1) Kelly and Amy had 50 jellybeans combined. Kelly gave Amy 12 of her jellybeans, then they had the same number of jellybeans. How many jellybeans did Kelly have to start with?

2) This puzzle contains the numbers 5-13. Each column and row add up to the number outside the boxes. Put the correct numbers in each box to complete the puzzle without repeating any numbers. What number should go in the top left box?

?	5		26
			27
10		7	28
30	29	22	

3) How many non-shaded rectangles of any size are in this shape?

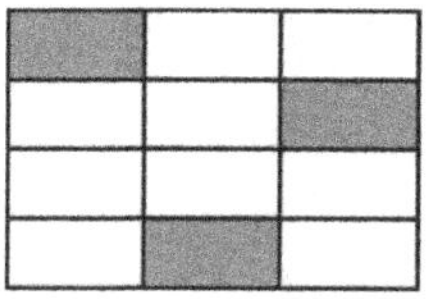

4) Albert and Bill have $32 combined. Bill and Craig have $30 combined. Albert and Craig have $34 combined. How much money does Craig have?

5) A farmer had 200 cows, but 60 of them died from anthrax. What fraction of his cows are still alive? (reduce/simplify your fraction)

6) Find the sum of the least and greatest prime number less than 40.

7) There are 90 students in the sixth grade. 50 like pizza, 46 like tacos, and 15 like both. How many students in the sixth grade class like neither pizza nor tacos?

8) In this multiplication problem different letters stand for different digits. What is the greatest value that ACC can be?

$$\begin{array}{r} BC \\ \underline{\times\ 6} \\ ACC \end{array}$$

9) The sum of Owen and Andy's ages is 51 years. Owen's age is twice Andy's age. How old is Owen?

10) The value of a two-digit number is four times the sum of its digits. The units digit is 2 more than the tens digit. What is the greatest two-digit number that fits these requirements?

11) 12 + 23 + 36 + 56 + 44 + 64 + 77 + 88 = ?

12) What is the difference between 65% of 56 and 56% of 65?

13) What is one half of one third of 144?

14) What is the greatest common factor of 144 and 420?

15) If July 1 falls on Tuesday, what day will Sept 2 fall on?

INDIVIDUAL ANSWER DOCUMENT
(Grades 5-6)

SCORE

Team # __________ School ______________________________

Student Name ______________________________________

Student Number _______________

1. ________	6. ________	11. ________
2. ________	7. ________	12. ________
3. ________	8. ________	13. ________
4. ________	9. ________	14. ________
5. ________	10. ________	15. ________

PERENNIAL MATH

Grade 5-6 Individual Event Solutions

1) In order for both girls to have the same amount of jellybeans, they would have 25 each in the end. So Kelly started with 25 + 12 = 37 jellybeans, and Amy started with 25 – 12 = 13 jellybeans. Check this out by adding 37+ 12 = 50 total jellybeans. Kelly started with 37 jellybeans.

2) Complete the bottom row by adding 11 to get a sum of 28. Then complete the middle column by adding 13 to get a sum of 29. The numbers left to choose from are {6, 8, 9, 12}. After you arrange these numbers to get the sums of the columns and rows, 12 is the top left number.

12	**5**	9	**26**
8	13	6	**27**
10	11	**7**	**28**
30	**29**	**22**	

3) Make a chart and count the different rectangles. There are 22 total rectangles in the shape.

Type	How Many
Made of 1 shape	9
Made of 2 shapes	9
Made of 3 shapes	3
Made of 4 shapes	1
Total Rectangles	**22**

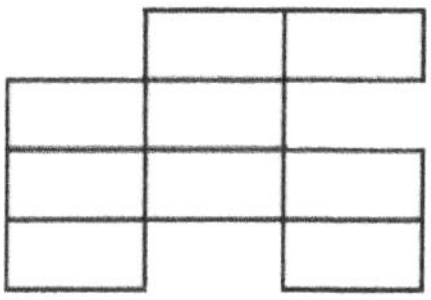

4) Use algebra. Let A = Albert, B = Bill, and C = Craig.

$$\begin{array}{rl} & A + B = 32 \\ & B + C = 30 \\ + & \underline{A + C = 34} \\ & 2A + 2B + 2C = 96 \end{array}$$

Divide both sides by 2 and A + B + C = 48. Substitute (A + B) in the equation to solve for C. 32 + C = 48. C equals 16. Craig has $16.

5) Subtract 200 – 60 = 140 cows. Reduce 140/200 to 7/10 of the cows remain.

6) The least prime number under 40 is 2 and the greater is 37. So, 2 + 37 = 39.

Grade 5-6 Individual Event Solutions

7) Use a Venn Diagram. The number of students who only like pizza is 50 - 15 = 35. The number of students who like only tacos is 46 - 15 = 31. To solve for neither subtract all the regions from the total. There are 9 students who like neither.

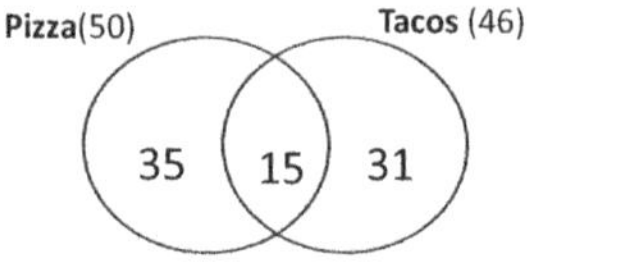

Neither 90 - (35 + 15 + 31) = 9

8) There are only 3 possible solutions for C (0, 4, or 8) since C x 6 = C. C cannot be 1, 2, 3, 4, 5, 6, 7, or 9 because the answer would carry and cannot be added to B x 6 to yield the same number C. C can be 0 and the product ACC would be 300. C could be 4 and the product would be 144. C could also be 8 and the product ACC could be 288 or 588. The greatest product ACC can be is 588.

BC	50	24	48	98
x 6	x 6	x 6	x 6	x 6
ACC	300	144	288	588

9) The sum of Owen and Andy's ages is 51 years. Owen's age is twice Andy's age. Let X stand for Andy's age. Then Owen's age is 2X. So, X + 2X = 51. 51/3 = X. X is 17 and 2X is 34. Owen is 34.

10) The units digit is 2 more than the tens digit. Possible numbers include: 13, 24, 35, 46, 57, 68, and 79. The value of a two-digit number is four times the sum of its digits, so the only number that fits this requirement is 24.

11) Use smart addition and group pairs together in sums of 100.

12 + 23 + 36 + 56 + 44 + 64 + 77 + 88 =
12 + 88 + 23 + 77 + 36 + 64 + 56 + 44 =
100 + 100 + 100 + 100 = 400

12) 65% of 56 = .65 x 56 = 36.40
56% of 65 = .56 x 65 = 36.40
The difference is zero.

13) One half of one third of 144 is 144/3 = 48, and then 48/2 = 24

14) Do the factorization of 144 {2 x 2 x 2 x 2 x 3 x 3} and the factorization of 420 {2 x 2 x 3 x 5 x 7}. The GCF is 2 x 2 x 3 = 12.

15) Both July and August have 31 days each. 31 + 31 = 62 days + 1 day in September. This would be 63 days which is a perfect 9 weeks. September 2 will be on a Tuesday.

INDIVIDUAL ANSWER KEY

(Grades 5-6)

Instructions for Scoring:

- ***Mark CORRECT answers by circling "O" the question number, and INCORRECT answers with an "X". Each correct answer is worth one point. Count the circles (correct answers) and write that number in the "SCORE" box.***

CORRECT → (1.) ________

INCORRECT → X. ________

1. 37
2. 12
3. 22
4. 16
5. 7/10
6. 39
7. 9
8. 588
9. 34
10. 24
11. 400
12. 0
13. 24
14. 12
15. Tuesday

PERENNIAL MATH

Grade 5-6 Individual Event (30 Minutes)

1) Today is Tuesday. What day of the week is it 180 days from today?

2) Calculate the area of the gray region. Use $\pi = 3.14$ and round to 2 decimal places.

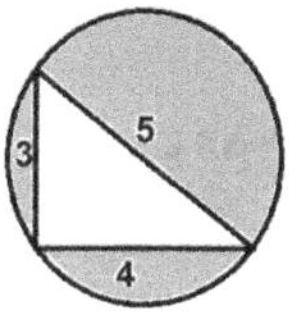

3) Complete the equation. What is the sum of the two missing numbers in the boxes?

$$7 = \frac{42}{\square} = \frac{\square}{15}$$

4) One banana plus one orange weighs the same as 14 strawberries. Two oranges weigh the same as 10 strawberries. How much does a banana weigh compared to strawberries?

5) Consecutive numbers are natural numbers that are in order such as 1, 2, and 3. Find the first of the seven consecutive numbers whose average is 15.

6) Mrs. Martin's class completed a math test. The average score of girls is 82, and the average score of the boys is 85. There are 67 girls and 55 boys in her class. What is the average math test score of the entire class? (round to 2 decimal points)

7) Max rides the subway twice a day for five days each week. Over a four week period, how much does he save by buying a $72 monthly pass compared to buying a $2.75 one-ride ticket each time he rides the subway?

8) Three boys divide 360 marbles equally among themselves. However, a fourth boy comes along and they put the marbles back in the pile and equally divide the marbles again among all four boys. How many more marbles did each boy receive when there were only three boys?

9) It is 3 a.m. Eastern time now in Atlanta, Georgia. In fifty-four hours Gretchen will be landing in Portland, Oregon. What time will the clocks say it is in Portland when she lands and she is on Pacific time then? Include a.m. or p.m. in your answer.

10) My piggybank contains only pennies, nickels, and dimes and the same number of each coin. The coins total $1.44. How many coins are there?

11) $2 + 6 \times 4 - 8 \times 2 =$

12) $702.9789 \div 78.37 =$

13) $6\frac{5}{6} \div 4\frac{3}{4} =$
(express as a reduced mixed fraction)

14) Find the GCF of 102 and 153.

15) $24 + 33 + 42 + 58 + 67 + 76 =$

INDIVIDUAL ANSWER DOCUMENT (Grades 5-6)

SCORE

Team # __________ School ______________________________

Student Name ______________________________________

Student Number ________________

1. __________	6. __________	11. __________
2. __________	7. __________	12. __________
3. __________	8. __________	13. __________
4. __________	9. __________	14. __________
5. __________	10. __________	15. __________

Grade 5-6 Individual Event Solutions

1) 180 ÷ 7 = 25 weeks with 5 days remainder. 5 days from Tuesday is Sunday.

2) The area of the shaded region is the area of the circle minus the area of the triangle. The radius is half the diameter or 5/2 = 2.5. (3.14 * 2.5 * 2.5) – (1/2 * 3 * 4) = 19.625 – 6 = 13.625. 13.625 rounded to two decimal places is 13.63.

3) Fill the missing numbers so that all equations equal 7. Add the two missing numbers together: 6 + 105 = 111.

$$7 = \frac{42}{\boxed{6}} = \frac{\boxed{105}}{15}$$

4) Since 2 oranges weigh the same as 10 strawberries, 1 orange weighs the same as 5 strawberries. Therefore, 1 banana's weight is the same as 14 – 5 = 9 strawberries.

5) To find the first of seven consecutive numbers whose average is 15, start with 15 in the middle because it is the average. The set of numbers is 12, 13, 14, 15, 16, 17, and 18. The first of these numbers is 12.

6) Find the total score from all the students: (82 x 67) + (85 x 55) = (5494 + 4675) = 10169
Divide the total score by the total number of students (67 + 55 = 122):
(82 x 67 + 85 x 55) ÷ (67 + 55) = (5494 + 4675) ÷ 122 = 83.35

7) For 4 weeks he would have to buy 40 individual passes. He would spend 40 x $2.75 = $110. He would save $110 - $72 = $38.

8) 360 marbles divided among 3 boys is 120 marbles each. 360 marbles divided among 4 boys is 90 marbles each. 120 – 90 = 30 more marbles for each of the 3 boys.

9) If it 3 a.m. Eastern time, then it is 12 a.m. Pacific time in Portland, Oregon. Fifty-four hours from then would be 2 full days plus 6 additional hours. 12 a.m. plus 6 hours would be 6 a.m.

Grade 5-6 Individual Event Solutions

10) Use algebra. The value of the coins is the number of cents for each coin multiplied by the number of that type of coin, so:

value of pennies: 1p
value of nickels: 5p
value of dimes: 10p

The total value is $1.44, so add the above set to equal 144 cents, and solve:

$$1p + 5p + 10p = 144$$
$$16p = 144$$
$$p = 9$$

There are nine of each type of coin in the piggy bank. There are 27 coins in all.

11) Use the order of operations PEMDAS to solve.

$2 + 6 \times 4 - 8 \times 2 =$

$2 + (24) - (16) = 10$

12) $702.9789 \div 78.37 = 8.97$

13) $6\frac{5}{6} \div 4\frac{3}{4} = \frac{41}{6} \div \frac{19}{4} = \frac{41}{6} \times \frac{4}{19} = \frac{164}{114} = \frac{82}{57} = 1\frac{25}{27}$

14) To find the GCF of 102 and 153 factor both numbers: 51 is the GCF.

```
Prime Factors Tree of 102        Prime Factors Tree of 153
      102                              153
      / \                              / \
     2   51                           3   51
    /      \                         /      \
   3        17                      3        17
  /           \                    /           \
 17            1                  17            1
```

15) Use smart addition and group pairs together.

$24 + 33 + 42 + 58 + 67 + 76 =$

$(24 + 76) + (33 + 67) + (42 + 58) =$

$100 + 100 + 100 = 300$

INDIVIDUAL ANSWER KEY
(Grades 5-6)

Instructions for Scoring:

- ***Mark CORRECT answers by circling "O" the question number, and INCORRECT answers with an "X". Each correct answer is worth one point. Count the circles (correct answers) and write the number in the "SCORE" box.***

CORRECT → (1.) ________

INCORRECT → X. ________

1. **Sunday**	6. **83.35**	11. **10**
2. **13.63**	7. **38**	12. **8.97**
3. **111**	8. **30**	13. $1\frac{25}{57}$
4. **9**	9. **6 a.m.**	14. **51**
5. **12**	10. **27**	15. **300**

PERENNIAL MATH

Grade 5-6 Individual Event (30 Minutes)

1) What is the difference between the largest and smallest prime number less than 50?

2) Each hexagon is made by adding the 2 hexagons below it. What number will be at the top?

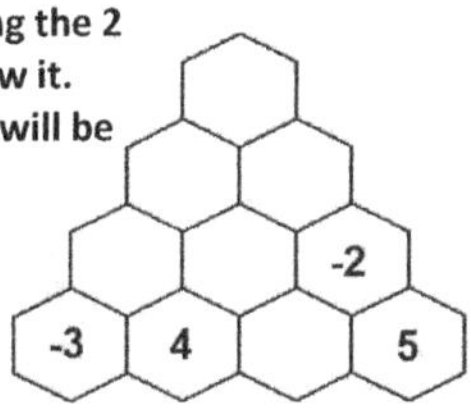

3) How many rectangles of any size are in this shape?

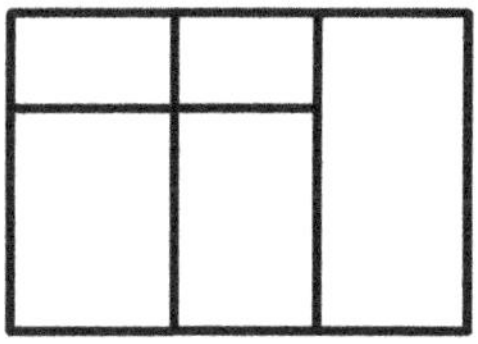

4) Consecutive numbers are numbers that are in order like 1, 2, 3. Find the three consecutive numbers whose sum is 93. What is the highest number in the set?

5) Ted's family ate 3/7 of the birthday cake. The next day his brother ate half of what was left. Then his sister ate half of what was left. Today, Ted got to finish his cake. How much cake did Ted get to eat today?

6) Janna has more than 150 coins and she can put them in equal stacks of 3, 4, and 7. What is the least number of coins she can have?

7) Megan and Patrick went to lunch at a cafe. They ordered a spinach salad for \$5.00, a tuna sandwich for \$4.85, and 2 glasses of lemonade for \$0.80 each. The tax was \$1.15. They gave the waiter \$15.00. How much change should they have received.

8) In this problem, each letter represents a different digit. What is the value of CDE?

$$\begin{array}{r} CC \\ +\ DD \\ EE \\ \hline CDE \end{array}$$

9) The carpenter made some 3-legged stools and some 4-legged chairs. When he finished there were 19 pieces of furniture and a total of 65 legs. How many chairs were there?

10) What is the median of the following set: (13, 27, 32, 29, 17, 34, 21).

11) 1 + 2 + 3 + 4 ... + 20 =

12) 2000
  x 555

13) 3.025 x 22.40 (do not round)

14) What is the least common multiple of 12 and 18 that is greater than 50?

15) 6 x 4 - 2 x 8 =

INDIVIDUAL ANSWER DOCUMENT
(Grades 5-6)

SCORE

Team # __________ School ________________________________

Student Name ____

Student Number ________________

1. __________	6. __________	11. __________
2. __________	7. __________	12. __________
3. __________	8. __________	13. __________
4. __________	9. __________	14. __________
5. __________	10. __________	15. __________

PERENNIAL MATH

Grade 5-6 Individual Event Solutions

1) What is the difference between the largest and smallest prime number less than 50? 47 is the greatest prime number 50 and 2 is the smallest. So, 47 – 2 = 45.

2) The top number is -7.

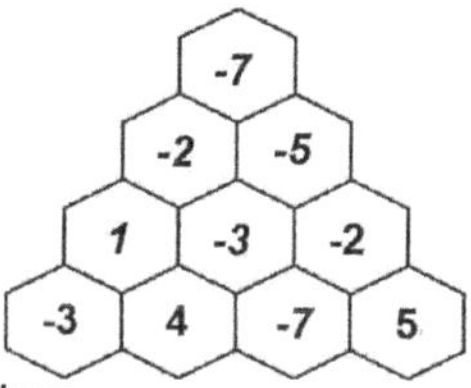

3) There are 12 rectangles in the shape.

Type	How Many
made from 1 shape	5
made from 2 shapes	4
made from 3 shapes	1
made from 4 shapes	1
made from 5 shapes	1
Total	12

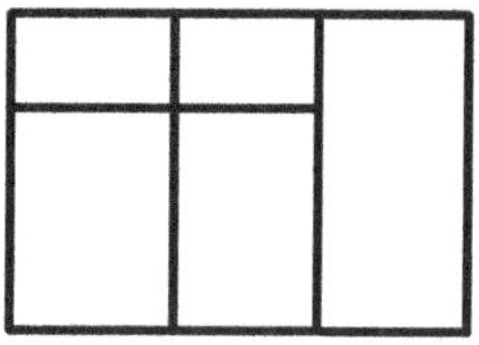

4) Use algebra. Three consecutive integers can be defined as: x, x + 1, and x + 2.

$$x + x + 1 + x + 2 = 93$$
$$3x + 3 = 93$$
$$3x = 90$$
$$x = 30$$

The three consecutive integers are 30, 31, and 32. 32 is the highest number in the set.

5) Ted's family ate 3/7 of the cake leaving 4/7. Now solve the remaining actions.

$$\frac{4}{7} x \frac{1}{2} = \frac{4}{14} \quad \frac{4}{14} x \frac{1}{2} = \frac{4}{28} = \frac{1}{7}$$

Ted ate 1/7 of the cake.

6) Janna has more than 150 coins and she can put them in equal stacks of 3, 4, and 7. To find the least number of coins she can have find the least common multiple. 3 and 7 are prime and 4 is the only other factor which as factors of 4, 2 and 1. Hence, the least common multiple of 3, 4 and 7 is 3 × 4 × 7 = 84. The least number of coins she could have is 168.

7) Add up the items they purchased and subtract that total from $15. They ordered a spinach salad for $5.00, a tuna sandwich for $4.85, and 2 glasses of lemonade for $0.80 each. The tax was $1.15. ($5.00 + $4.85 + $1.60 + $1.15) = $12.60. They gave the waiter $15.00. $15.00 - $12.60 = $2.40 change.

Grade 5-6 Individual Event Solutions

8) What is the value of CDE? Since C + D + E yields an E in the units place, then C + D must add up to 10. Since the answer is a three-digit number, then C has to be 1 or 2. If C is 1 then D is 9. E would be 8.

```
   CC
 + DD
   EE
  ---
  CDE
```

9) The carpenter made some 3-legged stools and some 4-legged chairs. When he finished there were 19 pieces of furniture and a total of 65 legs. To find out how many chairs there were make a chart. There were 8 chairs.

Chairs	10	9	8
Stools	9	10	11
Chair Legs (x4)	40	36	32
Stool Legs (x3)	27	30	33
Total Legs	67	66	65

10) What is the median of the following set: (13, 27, 32, 29, 17, 34, 21). The "median" is the "middle" value in the list of numbers. Arrange the list in order: 13, 17, 21, 27, 29, 32, 34. The middle number in this list is 27.

11) 1 + 2 + 3 + 4 ... + 20 =
You could list all the numbers out and add them up or use the formula S= n(n+1)/2 where S stands for the sum and n stands for the number of numbers in the sequence.
S = 20(20+1)/2 =
20(21)/2 = 420/2 =210

12)
```
     2000
    x 555
    10000
   100000
  1000000
  1110000
```

13) 3.025 x 22.40 = 67.76

14) What is the least common multiple of 12 and 18 that is greater than 50? The LCM of 12 and 18 is 36 so the next multiple that would be greater than 50 is 72.

15) Use the order of operations PEMDAS to solve.
6 x 4 - 2 x 8 =
24 - 16 = 8

PERENNIAL MATH

INDIVIDUAL ANSWER KEY

(Grades 5-6)

Instructions for Scoring:

- *Mark CORRECT answers by circling "O" the question number, and INCORRECT answers with an "X". Each correct answer is worth one point. Count the circles (correct answers) and write that number in the "SCORE" box.*

CORRECT → (1.) ________

INCORRECT → X. ________

1. 45
2. -7
3. 12
4. 32
5. 1/7
6. 168
7. 2.40
8. 198
9. 8
10. 27
11. 210
12. 1,110,000
13. 67.76
14. 72
15. 8

PERENNIAL MATH

Grade 5-6 Individual Event (30 Minutes)

1) There are 46 children at the zoo. Twenty-nine students like popcorn and twenty-five like peanuts. Fifteen like both. How many students like neither popcorn nor peanuts?

2) These cubes are glued together and dropped in a bucket of red paint. How many cubes have red paint on only 5 sides?

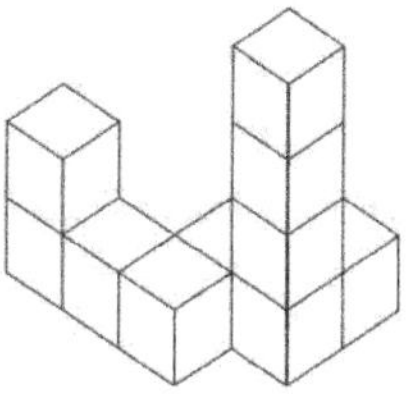

3) How many triangles of any size are in this shape?

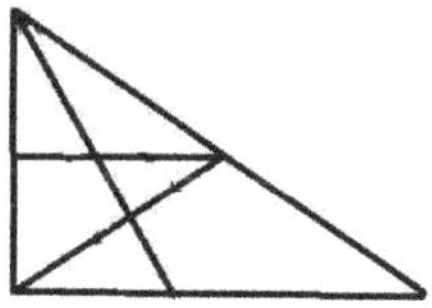

4) The zookeepers loaded some deer and emus on the train. There were 32 heads and 84 legs in all. How many deer were on the train?

5) A book is open. The sum of the two open page numbers is 377. What is the next page number in the book?

6) What is the sum of the first 5 prime numbers?

7) Miley's secret code is a four-digit multiple of 9 with no digits repeated. The thousands digit is the sum of the tens digit and the ones digit. The sum of all four digits is 18. What is largest number that Miley's secret code can be?

8) Mary and Glen have been married for thirty-five years. They dated for thirteen months before they were married. If today is January 18, 2018 what year did they start dating?

9) A book has 99 pages. How many digits did it take to number the 99 pages?

10) Daylee had $75. She bought 3 dozen doughnuts for $11.99 each, a gallon of orange juice for $5.79, and a gallon of milk for $3.59. How much money did she have left?

11) 13 + 36 + 57 + 43 + 64 + 87 =

12) 555
 x 22

13) $2 \times (4 + 3)^2$ =

14) What is the ones digit in the product of 5 x 10 x 20 x 25 x 30 x 35 ?

15) 94686 ÷ 367 =

INDIVIDUAL ANSWER DOCUMENT (Grades 5-6)

SCORE

Team # __________ School ______________________________

Student Name ______________________________________

Student Number ________________

1. ________	6. ________	11. ________
2. ________	7. ________	12. ________
3. ________	8. ________	13. ________
4. ________	9. ________	14. ________
5. ________	10. ________	15. ________

PERENNIAL MATH

Grade 5-6 Individual Event Solutions

1) Use a Venn Diagram. The number of students who only like popcorn = 29 - 15 = 14. The number of students who only like peanuts is 25 - 15 = 10. To solve for neither, subtract all the regions from the total. There are 7 students who like neither.

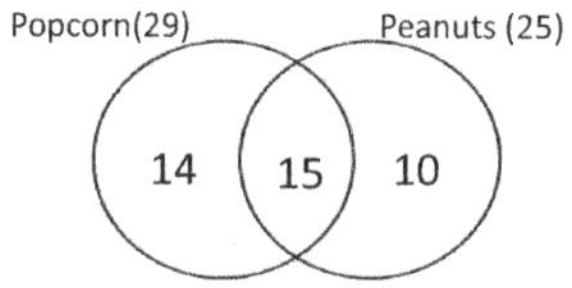

Neither 46 - (14 + 15 + 10) = **7**

2) The shape is made of 10 cubes. The cubes that have 1 block touching them have 5 sides showing. There are 3 cubes that have 5 red sides.

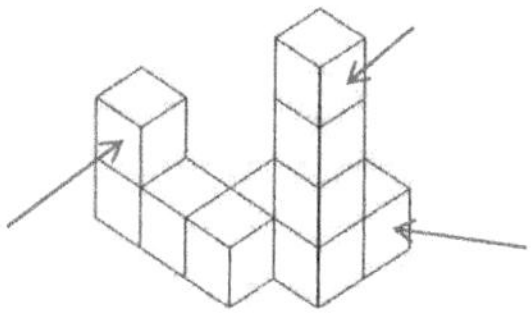

3) There are 15 triangles in the shape.

Type	How Many
made from 1 shape	6
made from 2 shapes	5
made from 3 shapes	2
made from 4 shapes	1
made from 6 shapes	1
Total triangles	15

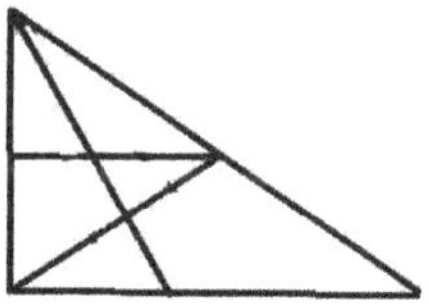

4) Make a table and start with 16 of each animal. You will notice that the number of legs decrease by 2 when you add an emu and take away a deer. There are 10 deer on the train.

Emus	**16**	**17**	**18**	**22**
Deer	**16**	**15**	**14**	**10**
Heads (32)	**32**	**32**	**32**	**32**
Legs (84)	**96**	**94**	**92**	**84**

5) 188 + 189 = 377. The next page is 190.

6) 2 + 3 + 5 + 7 + 11 = 28

7) The largest 4-digit number that is a multiple of 9, whose thousands digit is the sum of the tens and ones digit and the total sum of all four digits is 18 is 9081.

Grade 5-6 Individual Event Solutions

8) Mary and Glen have been married for thirty-five years. They dated for thirteen months before they were married. If today is January 18, 2018 what year did they start dating? 2018 – 35 years = 1983. If they got married in January of 1983 and started dating thirteen months prior, then they would have had to be dating by December 1981.

9) Count the page 1-9 and there are 9 digits. Then pages 10-19, 20-29, 30-39, 40-49, 50-59, 60-69, 70-79, 80-89, 90-99 there are 20 digits each for a total of 180 digits. 9 digits + 180 digits = 189 digits. The total number of digits in the 99 page book is 189.

10) Add up all the items and subtract from $75. The doughnuts cost $11.99 x 3 = $35.97. Add all the items she bought now: $35.97 + $5.79 + $3.59 = $45.35. $75 - $45.35 = $29.65.

11) Use smart addition and group pairs together.

13 + 36 + 57 + 43 + 64 + 87 =
13 + 87 + 36 + 64 + 57 + 43 =
100 + 100 + 100 = 300

12) 555 x 22 = 12,210

13) Use order of operations: PEMDAS

$2 \times (4 + 3)^2 =$
$2 \times (7)^2 =$
$2 \times 49 = 98$

14) What is the units digit in the product 5 x 10 x 20 x 25 x 30 x 35 = ?
Since there is a 0 in the units digit during the multiplication problem, the units digit will always end in 0. The answer to the multiplication problem is 26,250,000 for those students who do the entire problem. There is a zero in the units digit.

15) 94686 ÷ 367 = 258

INDIVIDUAL ANSWER KEY

(Grades 5-6)

Instructions for Scoring:

- *Mark CORRECT answers by circling "O" the question number, and INCORRECT answers with an "X" . Each correct answer is worth one point. Count the circles (correct answers) and write that number in the "SCORE" box.*

CORRECT → (1.) ______

INCORRECT → X. ______

1. 7
2. 3
3. 15
4. 10
5. 190
6. 28
7. 9,081
8. 1981
9. 189
10. 29.65
11. 300
12. 12,210
13. 98
14. 0
15. 258

PERENNIAL MATH

TEAM TESTS
GRADES 5-6

20 minute practice tests for teams
Each member gets a copy of the test, but the team only receives one answer document

Grade 5-6 Team Event (20 Minutes)

1) Hayden scored 142, 140, and 146 in his first 3 bowling games. What will he need to score in the fourth game to have a 148 average?

2) How many triangles are in this figure?

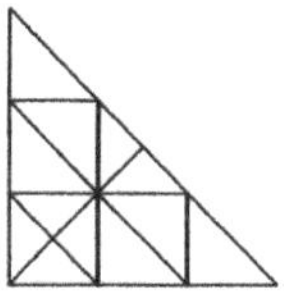

3) In this magic square, certain numbers can be place in the empty boxes so that the sum of the three numbers in each row, column or diagonal are equal. What number should be in the center box?

9		17
5	15	

4) Suzy recorded her tips at breakfast, lunch, and dinner. She made twice as many tips at lunch than breakfast. She made three times as many tips at dinner than breakfast. Altogether she made $576. How much did she make at dinner?

5) Jack was born at 7:20 a.m. on September 2, 2016. Jim was born on December 2, 2016 at 7:20 a.m. How many hours older is Jack than Jim?

6) In the theater there are 15 rows of seats. Each row has the same number of seats. If we were to take away 3 rows and move those seats to the remaining 12 rows, then the theater could hold the same number of people but the remaining 12 rows would have 3 extra seats. How many seats are in the theater?

7) If Kevin can wash a car in 48 minutes and Darren can wash a car in 16 minutes, how many minutes will it take them to wash a car working together?

8) Andrew and Brian have 29 dollars together. Brian and Clarence have 31 dollars combined. And Andrew and Clarence have 28 dollars together. How many dollars does Clarence have?

9) What is the smallest number that is divisible by 3 and 5 and leaves a remainder of 3 when it is divided by 8?

10) Brenda is reading a book with 250 pages. How many digits were used to number all the pages of the book?

PERENNIAL MATH

TEAM ANSWER DOCUMENT

SCORE

(Grades 5-6)

Team # __________ School ______________________________________

Student Names____

____

Write answers clearly. Each correct answer is worth one point.

1. __________
2. __________
3. __________
4. __________
5. __________
6. __________
7. __________
8. __________
9. __________
10. __________

PERENNIAL MATH

Grade 5-6 Team Event Solutions

1) Hayden scored 142, 140, and 146 in his first 3 bowling games. So his total sum was 428. For him to have an average of 148 in 4 games, his sum would need to be 4 x 148 = 592. 592 – 428 = 164. He will need to bowl 164 in the fourth game.

2) There are 25 triangles.

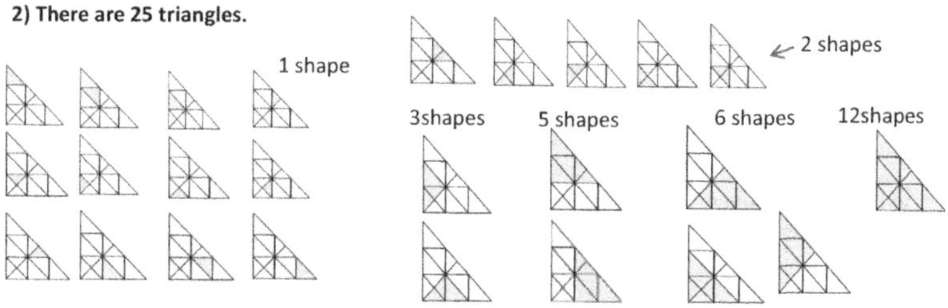

3) Look for shared squares to start with. Since the bottom right square is a shared square, you can solve for the missing number in the box above it. 5 + 15 = 17 + x, so the x = 3. Next you can use the middle as a shared square and solve for the bottom right box. 5 + 17 = 9 + x, so the bottom right is 13. Now solve for the sum 17 + 3 + 13 = 33. So the middle box is 9 + x + 13 = 33. x = 11 The middle box is 11.

9		17
		3
5	15	13

4) Suzy recorded her tips at breakfast, lunch, and supper. She made twice as many tips at lunch and three times as many tips at supper. Altogether she made $576. Use algebra. She made x at breakfast, 2x at lunch and 3x at supper. x + 2x + 3x = 576. 6x = 576. x = 96. Therefore she made 3x or $288 at supper.

5) Jack was born at 7:20 a.m. on September 2, 2016. Jim was born on December 2, 2016 at 7:20 a.m. There are 28 more days in September, 31 in October, 30 in November and then the 2 days in December. 91 days x 24 hours = 2184 hours difference in age.

6) You could draw this out and have 15 rows. Since the remaining 12 rows are receiving 3 seats each, then there were 36 seates moved from those 3 rows. You can deduce that each row had 12 seats to start with since you moved 36 seats and they made up 3 rows. 15 original rows times 12 seats equal 180 seats in the theater.

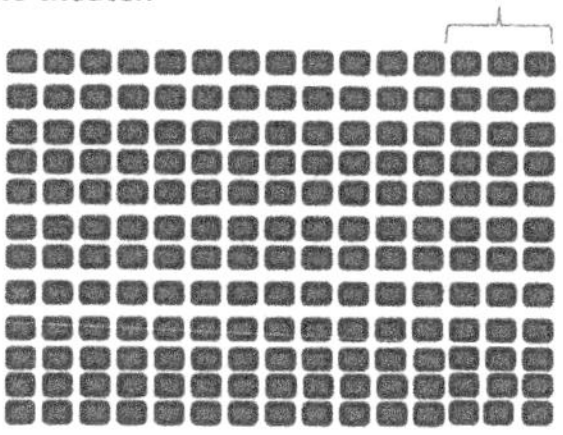

These 3 rows will move and add 3 seats to the remaining 12 rows, so 3 x 12 = 36.

Since there were 3 rows moved 36/3 = 12 seats in each row at the start.

Grade 5-6 Team Event Solutions

7) If Kevin can wash a car in 48 minutes and Darren can wash a car in 16 minutes, how many minutes will it take them to wash a car working together? Use algebra. Kevin can do the job in 1/48 and Darren can wash the car in 1/16.

$$1/48 + 1/16 = 1/t$$
$$1/48 + 3/48 = 1/t$$
$$4/48 = 1/t$$
$$1/12 = 1/t$$
$$12/1 = t$$ It will take them 12 minutes to wash the car together.

8) Use Algebra with A = Andrew, B = Brian, and C = Clarence.

$$\begin{aligned} A + B \quad &= 29 \\ B + C &= 31 \\ \underline{A + \quad C} &= \underline{28} \\ 2A + 2B + 2C &= 88 \end{aligned}$$

Divide by 2 and A + B + C = 44. Now you can substitute the original equation (A + B) into this and deduce that 29 + C = 44, So C = 15; Clarence has $15

9) To find the smallest number that is divisible by 3 and 5 and leaves a remainder of 3 when it is divided by 8, first find the least common multiple of 3 and 5. The LCM of 3 and 5 is 15. So look at multiples of 15 to determine which one leaves a remainder of 3 when divided by 8. Multiples of 15: 15, 30, 45, 60, 75, 90. The smallest multiple that leaves a remainder of 3 when divided by 8 is 75. 75 is the number that meets the requirements.

10) Brenda is reading a book with 250 pages. To find out how many digits were used to number all the pages of the book, create charts to keep up with the digits. There were 642 digits used.

Numbers	Digits
1-10	11
11-20	20
21-30	20
31-40	20
41-50	20
51-60	20
61-70	20
71-80	20
81-90	20
91-100	21
	192

Numbers	Digits
101-110	30
111-120	30
121-130	30
131-140	30
141-150	30
151-160	30
161-170	30
171-180	30
181-190	30
190-200	30
201-210	30
211-220	30
221-230	30
231-240	30
241-250	30
	450

TEAM ANSWER KEY

(Grades 5-6)

__Instructions for Scoring:__

- ***Mark CORRECT answers by circling "O" the question number, and INCORRECT answers with an "X" . Each correct answer is worth one point. Count the circles (correct answers) and write that number in the "SCORE" box.***

CORRECT → (1.) ________

INCORRECT → X. ________

1. **164**
2. **25**
3. **11**
4. **288**
5. **2,184**
6. **180**
7. **12**
8. **15**
9. **75**
10. **642**

PERENNIAL MATH

Grade 5-6 Team Event (20 Minutes)

1. What is the sum of all the first 7 composite numbers?

2) This tower is made of several cubes. Some cubes cannot be seen in the picture. How many cubes in all did it take to make this tower?

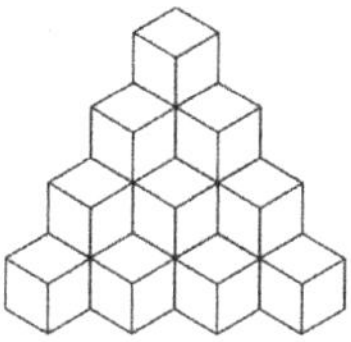

3) How many rectangles of any size are in this diagram?

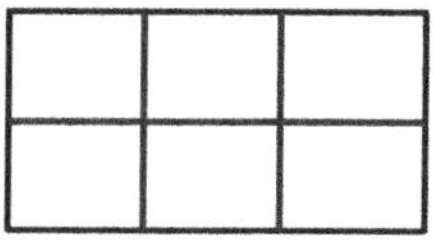

4) How many rectangles can be made with an area of $48cm^2$ if the length of the sides are whole numbers?

5) There are 46 children at the park. Twenty-nine students like hot dogs and twenty-five like hamburgers. Fifteen like both. How many students like neither hotdogs nor hamburgers?

6) Ashley, Bonita and Carly go shopping. When they pull out their money they discover that Ashley and Bonita have \$27 together. Bonita and Carly have \$31 combined. Ashley and Carly have \$34 together. Who has the most money and how much money does she have?

7) Eve was elected class president. She received 3 votes for every 2 votes that Donald got. No one else ran. However, if 6 of the people who voted for Eve had voted for Donald instead, Eve would have received 2 votes for every 3 that Donald would have received. How many students voted?

8) In this subtraction problem A and B stand for different digits. What is the value of B – A ?

$$\begin{array}{r} 5A \\ -\ \underline{2B} \\ 28 \end{array}$$

9) Jonah went to a store and spent one-third of his money. He went to a second store where he spent one-third of what remained, and then had \$16 when he left. How much money did he have to begin with at the first store?

10) A rectangular swimming pool measures 12 meters wide by 18 meters long. It is surrounded by a sidewalk that is 4 meters wide around the whole outside perimeter. What is the area of the sidewalk?

TEAM ANSWER DOCUMENT

(Grades 5-6)

SCORE

Team # __________ School ________________________________

Student Names____

____

Write answers clearly. Each correct answer is worth one point.

1. ___________

2. ___________

3. ___________

4. ___________

5. ___________

6. ___________

7. ___________

8. ___________

9. ___________

10. ___________

PERENNIAL MATH

Grade 5-6 Team Event Solutions

1) The sum of first seven composite numbers is 4 + 6 + 8 + 9 + 10 + 12 + 14 = 63

2) This tower is made up of several cubes. Count them by layer.

- Bottom layer is composed of 4 + 3 + 2 +1 = 10.
- 2nd layer is 3 + 2 + 1= 6
- 3rd layer is 2 + 1 = 3
- Top layer is 1

10 + 6 + 3 + 1 = 20 cubes total.

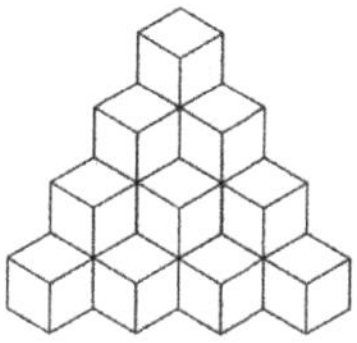

3) How many rectangles of any size are in this diagram?

Type	How Many
Made of 1 shape	6
Made of 2 shapes	7
Made of 3 shapes	2
Made of 4 shapes	2
Made of 6 shapes	1
Total	18

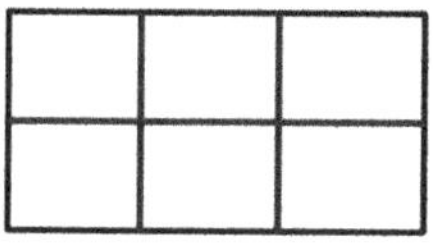

4) Make a chart to see the different rectangles: There are 5 rectangles with an area of 48cm^2.

Length	Width
48	1
24	2
16	3
12	4
8	6

5) Use a Venn Diagram. The number of students who only bought hotdogs = 29 - 15 = 14. The number of students who only bought hamburgers = 25 - 15 = 10. To solve for neither subtract all the regions from the total. There are 7 children who bought neither.

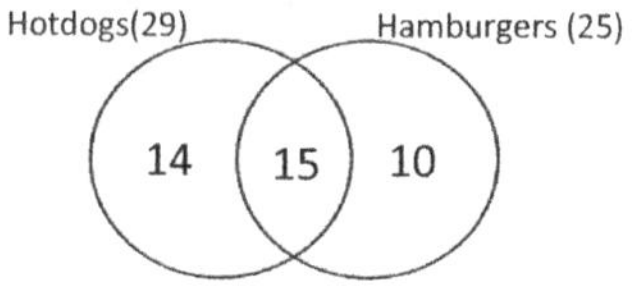

Neither 46 - (14 + 15 + 10) = **7**

6) Use Algebra. Let A = Ashley, B = Bonita, and C = Carly.

$$\begin{array}{lcl} A + B & & = 27 \\ & B + C & = 31 \\ \underline{A +} & \underline{\quad C} & \underline{= 34} \\ 2A + 2B + 2C & & = 92 \end{array}$$

Divide by 2 and A + B + C = 46. Now you can substitute the original equation (A + B = 27) so that 27 + C = 46. So C is <u>19</u>. Substitute C into B + C = 31 to get B + 19 = 31. Therefore B is <u>12</u>. Substitute 12 into A + B = 27 to get A + 12 = 27, therefore A is <u>15</u>. Carly has the most $19.

Grade 5-6 Team Event Solutions

7) Set up a table where the first scenario has a 3:2 ratio and the second scenario subtracts 6 votes from Eve and adds 6 votes to Donald. Look for numbers in the second scenario that have a 2:3 ratio. In the original election, Eve received 18 votes to Donald's 12 votes so there were 30 students who voted.

Eve's Votes	3	6	9	12	15	18	21	24
Donald's Votes	2	4	6	8	10	12	14	16
Eve's Votes - 6	-3	0	3	6	9	12	15	18
Donald's Votes + 6	8	10	12	14	16	18	20	22

8) There are several possible solutions to the subtraction problem: 51 – 23 = 28; 52 – 24 = 28; 53 – 25 = 28; 54 – 26 = 28, ... In all cases B must be greater than A. Therefore, B – A = 2

$$\begin{array}{r} 5A \\ -\ \underline{2B} \\ 28 \end{array}$$

9) Jonah spent 1/3 of his money in store 1 and had 2/3 left. In the second store he spent 1/3 of 2/3 of his original amount. In both stores he spent 1/3 + (1/3 x 2/3). 3/9 + 2/9 = 5/9 of his money. Therefore he had 4/9 of his money left which was equal to $16. Then 1/9 of his money was $4. He originally had 9 x 4 or $36.

	Amount When Entering Store	Spent 1/3	Left 2/3
First Store	36	12	24
Second Store	24	8	16

10) Draw the area. The entire area would be 26m x 20m = $520m^2$. Now find the area of the swimming pool which is 18m x 12m = $216m^2$. Subtract the area of the swimming pool from the entire area to get the area of the sidewalk: $520m^2$ - $216m^2$ = $304m^2$.

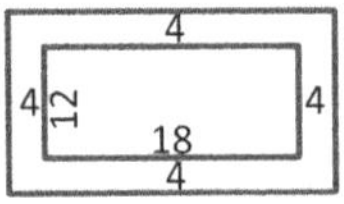

TEAM ANSWER KEY

(Grades 5-6)

Instructions for Scoring:

- ***Mark CORRECT answers by circling "O" the question number, and INCORRECT answers with an "X" . Each correct answer is worth one point. Count the circles (correct answers) and write that number in the "SCORE" box.***

CORRECT → (1.) ________

INCORRECT → X. ________

1. **63**
2. **20**
3. **18**
4. **5**
5. **7**
6. **Carly 19**
7. **30**
8. **2**
9. **36**
10. **304**

Grade 5-6 Team Event (20 Minutes)

1) The sum of three consecutive odd numbers is 117. What is the product of those three numbers?

2) In this magic square, the numbers 1-16 are placed in the boxes so that the sum of each row, column, and diagonal is the same amount. Fill in the missing numbers and find the sum of A + B.

	A	2	13
5		11	
B	6		12
4		14	

3) It takes Merle 6 minutes to cut the log into 3 pieces. How many minutes will it take him to cut it into 6 pieces at the same rate?

4) Eric has $14 more than Craig and $17 more than Trevor. Together the boys have $89. How much does Eric have?

5) How many numbers between 1-1000 are multiples of both 4 and 5 but not of 6?

6) If b ◇ (c ■ d) means (b + c) x (b + d), how much larger is 4 ◇ (5 ■ 6) than 3 ◇ (4 ■ 5)?

7) Out of 30 students in Mrs. Mitchell's class, 9 take art, 11 take music, and 4 take both subjects. How many students take neither art or music?

8) In this multiplication problem A, B, and C stand for different digits. What is the sum of A, B and C?

$$\begin{array}{r} BA \\ \underline{\times\ 9} \\ CAA \end{array}$$

9) 1) It costs 23¢ to mail a postcard and 35¢ to mail a letter. If Bernie spent $5.35 and wrote postcards and letters to 17 friends, how many letters did he mail?

10) The Smith family are going to the movies. It costs $9 for adult tickets and $6.50 for children. Grandpa paid $120 for the total cost of all 15 people. How many adult tickets were bought?

TEAM ANSWER DOCUMENT
(Grades 5-6)

SCORE

Team # __________ School ______________________________

Student Names______________________________________

____

Write answers clearly. Each correct answer is worth one point.

1. __________

2. __________

3. __________

4. __________

5. __________

6. __________

7. __________

8. __________

9. __________

10. __________

PERENNIAL MATH

Grade 5-6 Team Event Solutions

1) Use algebra to solve. The sum of the three consecutive numbers can be represented as x, x + 2, and x + 4.

$$x + x + 2 + x + 4 = 117$$
$$3x + 6 = 117$$
$$3x = 111$$
$$x = 37$$

If x = 37, then the other two numbers are 39 and 41. Their product 37 x 39 x 41 = 59,163.

2) The sum of the diagonal is 4 + 6 + 11 + 13 = 34.
Solve for x in the third column, x = 34 – (2 + 11 + 14) = 7.
Solve for B in the third row, B = 34 – (6 + 7 + 12) = 9.
Solve for y in the first column, y = 34 – (5 + 9 + 4) = 16.
Solve for A in the first row, A = 34 – (16 + 2 + 13) = 3.
The sum of A + B = 3 + 9 = 12.

y	*A*	2	13
5		11	
B	6	x	12
4		14	

3) It takes Merle 6 minutes to cut the log into 3 pieces which is ONLY 2 cuts. So it takes him 3 minutes to make a cut in the log. It will take 5 cuts to make 6 pieces. 5 x 3 = 15 minutes.

4) Use algebra to solve. Let X represent Eric's amount. Craig's amount is X - 14. Trevor's amount is X - 17.

$$X + (X - 14) + (X - 17) = 89$$
$$3X = 89 + 14 + 17$$
$$3X = 120$$
$$X = 120/3 = 40.$$ Eric has $40

5) Since 1000/(4 x 5) = 1000/20 = 50, 50 are multiples of 4 and 5. The LCM of 3, 4 and 5 is 60. Since 1000/60 = 16.666, there are also 16 multiples of 6; 50 - 16 = 34. Now you must also exclude 1000 because it stated the numbers between 1-1000. There are 33 multiples.

6) If b ◇ (c ■ d) means (b + c) x (b + d), how much larger is 4 ◇ (5 ■ 6) than 3 ◇ (4 ■ 5)?

(4 + 5) x (4 + 6) = 9 x 10 = 90   (3 + 4) x (3 + 5) = 7 x 8 = 56.

90 – 56 = 34. The difference is 34.

Grade 5-6 Team Event Solutions

7) Create a Venn Diagram. Total art is Region I + Region II. Total music is Region II + Region III. Neither art nor music is Region IV and can be found by subtracting the sum of Region I, II, and III from the total 30. 30 – (5 + 4 + 7) = 14. There are 14 students who do not take art or music.

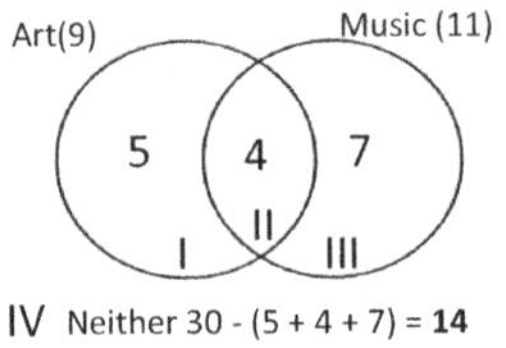

8) A must equal 5 or 0 for 9 x A to equal A. A cannot equal 0 because the answer would have to end in 0 for 9 x B. Since A = 5, Then 9 x B + 4 that carried = C5. B would have to be 9 and C would equal 8. The sum of the three numbers is 5 + 9 + 8 = 22.

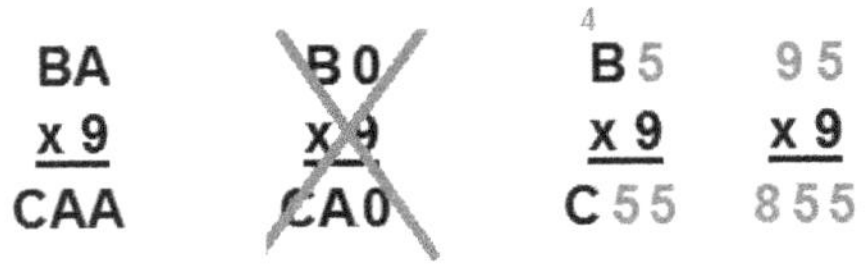

9) Make a table. Start with 8 postcards and 9 letters. Increase the number of letters until the cost is $5.35. He mailed 12 letters.

Postcards (0.23)	8	7	6	5
Letters (0.35)	9	10	11	**12**
Total Cost	4.99	5.11	5.23	5.35

10) Make a table to see the cost for 15 people with adults at $9 and children at $6.50. Change the number of adults and children until the total paid is $120. There were 9 adults.

Number of Adults	1	2	3	4	5	6	7	8	**9**
Number of Children	14	13	12	11	10	9	8	7	6
Adult Tickets (x $9)	9	18	27	36	45	54	63	72	81
Children Tickets (x $6.50)	91	84.5	78	71.5	65	58.5	52	45.5	39
Total Cost	100	102.5	105	107.5	110	112.5	115	117.5	120

TEAM ANSWER KEY

(Grades 5-6)

Instructions for Scoring:

- ***Mark CORRECT answers by circling "O" the question number, and INCORRECT answers with an "X" . Each correct answer is worth one point. Count the circles (correct answers) and write that number in the "SCORE" box.***

CORRECT ⟶ (1.) ________

INCORRECT ⟶ X. ________

1. **59,163**
2. **12**
3. **15**
4. **40**
5. **33**
6. **34**
7. **14**
8. **22**
9. **12**
10. **9**

PERENNIAL MATH

Grade 5-6 Team Event (20 Minutes)

1) What number is 2 less than the sum of the first 7 prime numbers?

2) Use the scale to solve the problem. How much does 1 apple, 1 orange, and 1 strawberry weigh altogether?

3) How many rectangles of any size are in this diagram?

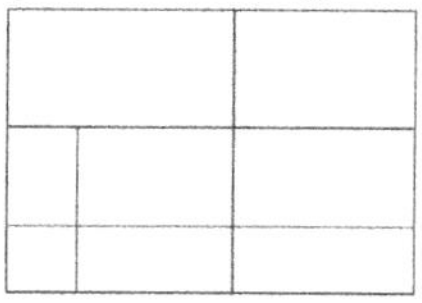

4) What is the perimeter of a rectangle with an area of $57cm^2$ and a width of 3cm?

5) In the weight room there are nine weights to choose from. There are at least one of each type: 20lb, 40lb and 50lb. There is a different number of each type of weight. What is the least total weight possible?

6) Jim is twice as old as Todd. Todd is three times as old as Bart. The sum of all three ages is 90 years. How old is Todd?

7) Aric, Bob, and Carl have money for the concert. Aric and Bob have $15 together. Bob and Carl have $16 together. Aric and Carl have $13 combined. How much money does Aric have?

8) In this magic square, five more numbers can be placed in the boxes so that the sum of the three numbers in each row, each column, and in each diagonal is the same amount. What is the value of N?

12		16
N		
x	18	8

9) Arnold and Ken each threw darts at a target. Arnold hit region *x* four times and region *y* two times for a score 42 points. Ken hit region *x* three times and region *y* three times for a score of 45 points. What was the value of region *x*?

10) Judy put her cupcakes in 16 equal rows. If she rearranges 4 rows, and moves those cupcakes so that there are 12 rows instead of 16, then each of those 12 rows would get 4 extra cupcakes. How many cupcakes does she have?

Grade 5-6 Team Event Solutions

1) The sum of first seven prime numbers is 2 + 3 + 5 + 7 + 11 + 13 + 17 = 58. 58 – 2 = 56.

2) Since we know that A + 2O weighs 36, we can subtract 36 from the 4 oranges and apple. 66 - 36 = 30. So, 2 oranges weigh 30 or 1 orange weighs 15. Now substitute 3 oranges x 15 = 45 in with the 5 strawberries. So, 65 - 45 = 20. 20/5= 4. So, one strawberry weighs 4. To solve for the apple, subtract 36 - 30 = 6. The weight of all three is A + O + S = 6 + 15 + 4 = 25.

3) How many rectangles of any size are in this diagram?

Type	How Many
Made of 1 shape	8
Made of 2 shapes	9
Made of 3 shapes	4
Made of 4 shapes	2
Made of 5 shapes	2
Made of 6 shapes	1
Made of 8 shapes	1
Total	**27**

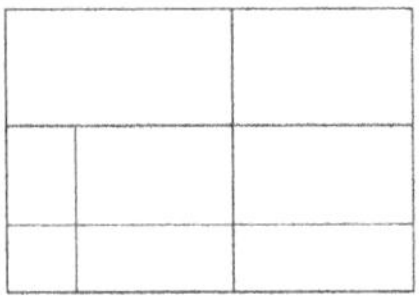

4) The area of a rectangle is length times width. Since the width is 3, the length must be 19 cm in order for the area to be 57cm^2. To get the perimeter add up the sides 3 + 19 + 3 + 19 = 44cm.

5) To find the least possible sum of the weights, first figure out the possible combinations of the type weights. There can be combinations of (1, 2, and 6); (1, 3, and 5); or (2, 3, and 4) weights to give you 9 weights with different amounts of each type. Now select fewer of the heaviest weights to give the lowest total weight. There is one 50lb weight, two 40lb weights, and six 20lb weights. Their total sum is 50 + 2(40) + 6(20) = 250lb.

 PERENNIAL MATH

Grade 5-6 Team Event Solutions

6) Make a chart to compare their ages: Todd is 27 years old.

Bart	Todd (3B)	Jim (2T)	Total (90)
5	15	30	50
6	18	36	60
7	21	42	70
8	24	48	80
9	**27**	54	90

7) Use Algebra. Let A = Aric, B = Bob, and C = Carl.

$$\begin{array}{rl} A + B \qquad & = 15 \\ B + C & = 16 \\ + \underline{A + \qquad C} & \underline{= 13} \\ 2A + 2B + 2C & = 44 \quad \text{divide both sides by 2} \\ A + B + C & = 22 \end{array}$$

Now you can substitute B + C = 16 into A + B + C = 22 to get A + 16 = 22. So, A is 6.

8) Use algebra to solve. Pick a corner that is shared by two sides. The bottom left corner is shared by the bottom row and the left column. So (8 + 18 + x) = (12 + N + x). Simplify and you get (8 + 18) = (12 + N). N is 14. The corner sums are equal, (8 + 12) = (16 + x). So, x is 4. Now you can solve for a side (8 + 18 + 4) = 30. The middle box is equal to (8 + m + 12) = 30. The middle box m is equal to 10.

12		16
N	m	
x	18	8

9) Use algebra to solve:
Arnold is 4x + 2y = 42.
Ken is 3x + 3y = 45 or divide both sides by 3 and you get x + y = 15. For convenience, multiply both sides by 2 for 2x + 2y = 30.
Now subtract Ken from Arnold.

$$\begin{array}{rl} 4x + 2y & = 42 \\ - \underline{2x + 2y} & \underline{= 30} \\ 2x \qquad & = 12 \\ & x = 6 \end{array}$$

10) You could draw this out and have 16 rows. Since you are moving 4 rows out to leave 12 rows and those 12 rows are each receiving 4 cupcakes added to them, then there were 48 cupcakes moved from those 4 rows. You can deduce that each row had to have had 12 in it to start with since you moved 48 cupcakes and they made up 4 rows. 16 original rows times 12 cupcakes equal 192 cupcakes.

PERENNIAL MATH

TEAM ANSWER KEY

(Grades 5-6)

Instructions for Scoring:

- *Mark CORRECT answers by circling "O" the question number, and INCORRECT answers with an "X". Each correct answer is worth one point. Count the circles (correct answers) and write that number in the "SCORE" box.*

CORRECT → (1.) ________

INCORRECT → X. ________

1. 56
2. 25
3. 27
4. 44
5. 250
6. 27
7. 6
8. 10
9. 6
10. 192

PERENNIAL MATH

Made in the USA
Monee, IL
05 November 2026

67811013R00036